Made in the USA
Middletown, DE
12 May 2022

65639743R00080